ABC
Cosmos

By Mastal

Asteroid - a small planetary body in orbit around the Sun. Most asteroids are located between the orbits of Mars and Jupiter.

Black hole -after its fuel is exhausted, the core of a very massive star will collapse under its own gravity, forming a black hole. The gravitational pull becomes so strong that not even light can escape.

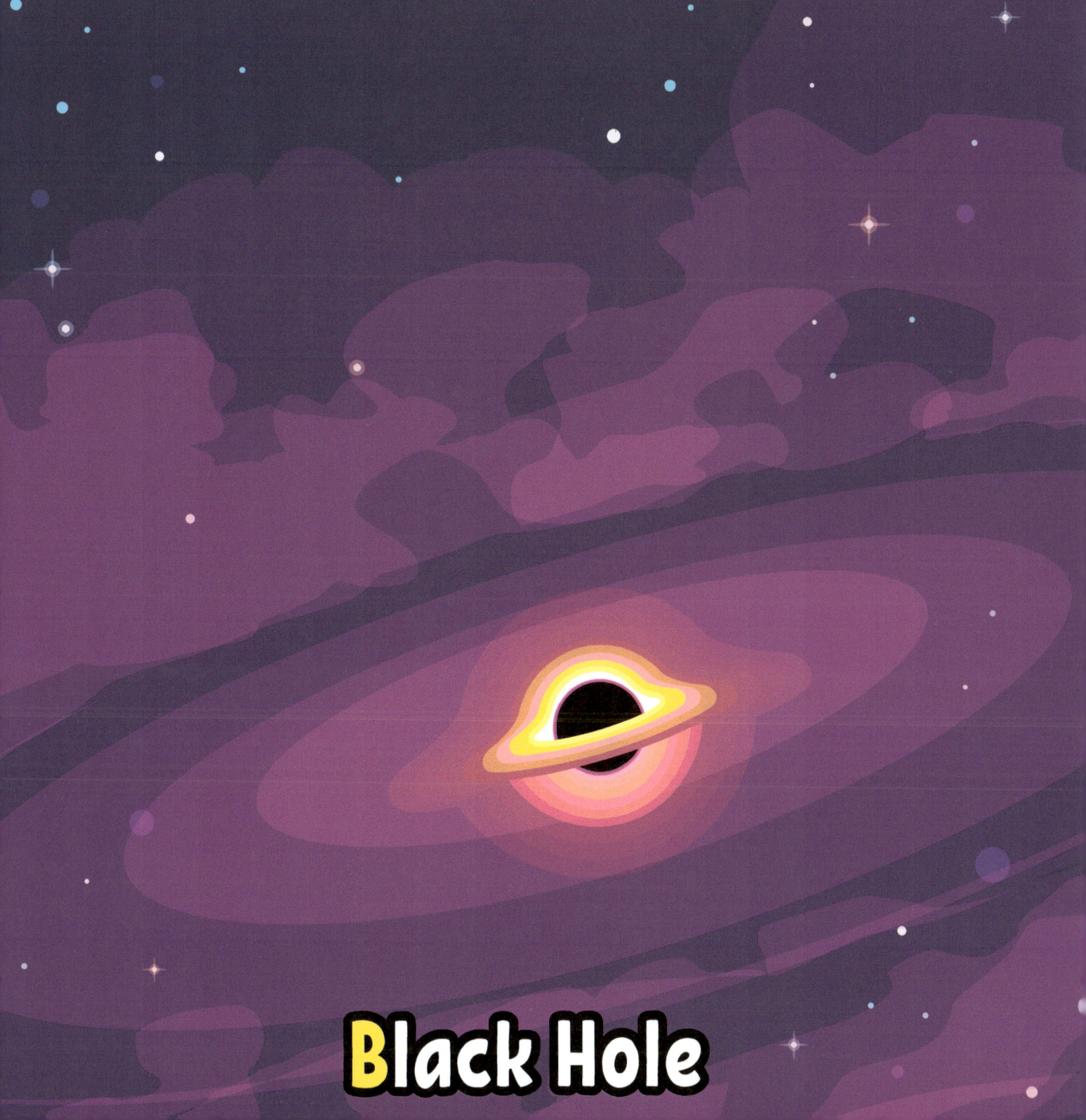

Comet- a celestial body of ice and rock that orbit the Sun forming a long tail of gas and dust when heated by the sun's rays.

Comet

Dwarf stars - a star, such as the sun, having relatively low mass, small in size, and has average or below average luminosity.

Dwarf Stars

Earth- the third planet from the sun, on which we live, and is the fifth largest planet in the solar system. Its name is the only planet that does not come from Greco-Roman mythology. Its meaning of ground and earth is drawn from Old English and Germanic words.

Earth

Fusion - a thermonuclear reaction in stars that require high temperatures and pressure which fuses hydrogen into helium.

Fusion

Galaxy - groupings of millions, billions, and trillions of stars held together by gravitational attraction, and found throughout the universe.

Hubble telescope- the Hubble telescope was launched into orbit, about 600 km above Earth, in 1990. It travels 5 miles per second and has taken pictures of planets, stars and galaxies all obtained through the visible, infrared, and ultraviolet ranges.

Hubble Telescope

Interstellar - region occurring between stars that contains vast, diffuse clouds of gases and minute solid particles.

Jupiter - the fifth planet from the sun. It is the largest of the planets and a gas giant. It was named after the Roman ruler of the gods and heavens.

Jupiter

Kuiper belt - a ring of small icy, primitive celestial bodies beyond the orbit of Neptune in which Pluto is a part of. It is named after the Dutch American astronomer Gerard P. Kuiper and comprises of hundreds of millions of objects.

Kuiper Belt

Lunar eclipse - an eclipse in which the moon appears darkened as it passes into Earth's shadow.

Lunar Eclipse

M

Mars - the fourth planet from the sun, and is the seventh in size and mass. Mars is red in color, and a terrestrial planet. Its named after the Roman god of war.

Neptune - the eighth planet from the sun and third most massive planet of the solar system. It is also a gas giant and is named after the Roman god of the sea.

Neptune

Orbit - the path of a celestial body as it moves through space around another celestial body, due to gravity and sideways motion.

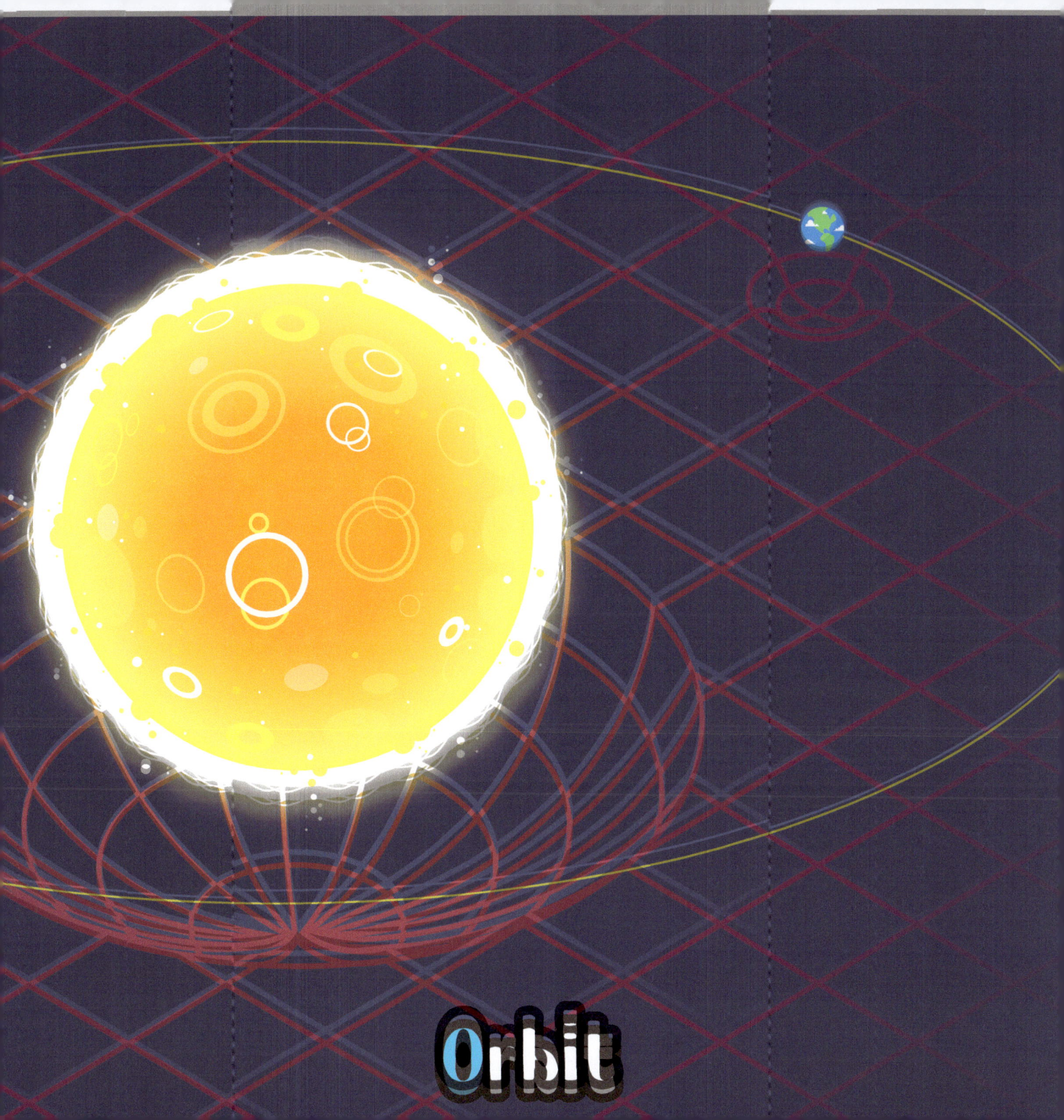

Pulsar - a spinning neutron star that emits energy along its gravitational axis. This energy is received as pulses as the star rotates.

Quasar- when gas spirals at high velocity into a black hole, but some are released back into space. They are among the oldest and brightest objects in the universe and believed to be the center of ancient, active galaxies.

Rocket - a vehicle designed to launch people and objects into space. This is accomplished by enclosing gas under pressure, in a chamber that has an opening allowing for gas to escape. This provides for thrust to propel the rocket.

Saturn - the sixth planet from the sun, and the second largest planet of the solar system in mass and size. It is also a gas giant, and was named after the Roman god of agriculture.

Saturn

Tides - the alternate rising and falling of the sea, due to the gravitational force of the moon which causes a bulge on the near side and another bulge on the far side. Earth turns within the bulges and results in the rise and fall of tides twice a day.

Tides

Uranus - the seventh planet from the sun, and a gas giant. It has methane in its Uranian atmosphere which absorbs red light from the sun, and reflects blue light giving the planet its blue-green color. It was named for its representation of heaven, the son, and husband of Gaea, in Greek mythology.

Uranus

Venus - the second planet from the sun, and the sixth largest planet in the solar system. It is also a terrestrial planet. It is the planet that comes closes to Earth and is the brightest of the planets when it is visible. It is named after the Roman goddess of love and beauty.

Venus

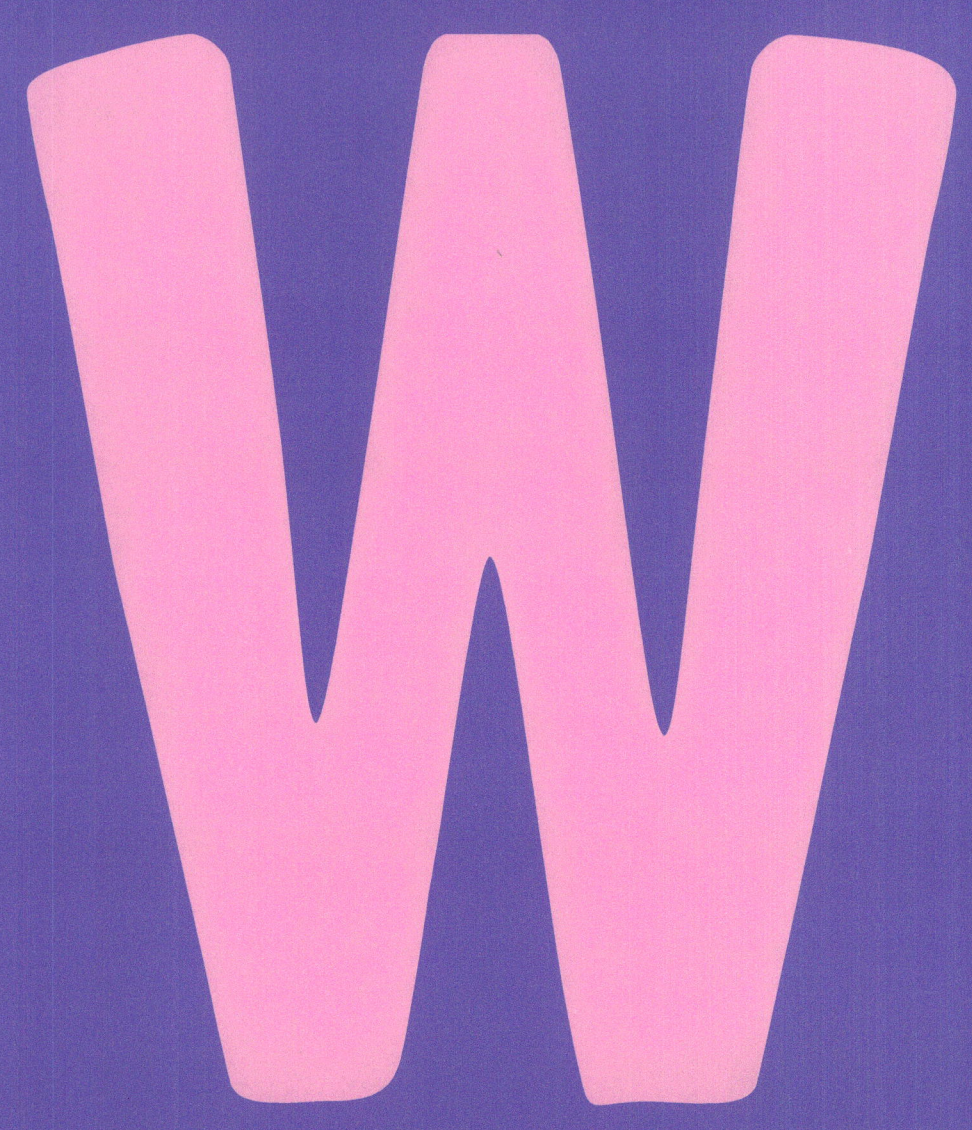

Wavelength - the distance between consecutive crests of a wave, which serves as a unit of measure of electromagnetic radiation.

Gamma Ray ~~~~~~~~~~~

X-Ray ~~~~~~~~~~

UV ~~~~~~~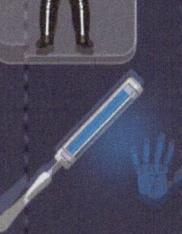

Visible Light ~~~~~

Infrared ~~~

Microwaves ~~

Wavelength
Wavelength

Radio Waves ~

Wavelength

X-ray - electromagnetic radiation of a very short wavelength. It is very high-energy and used in photographic or digital imaging of the internal composition of the body.

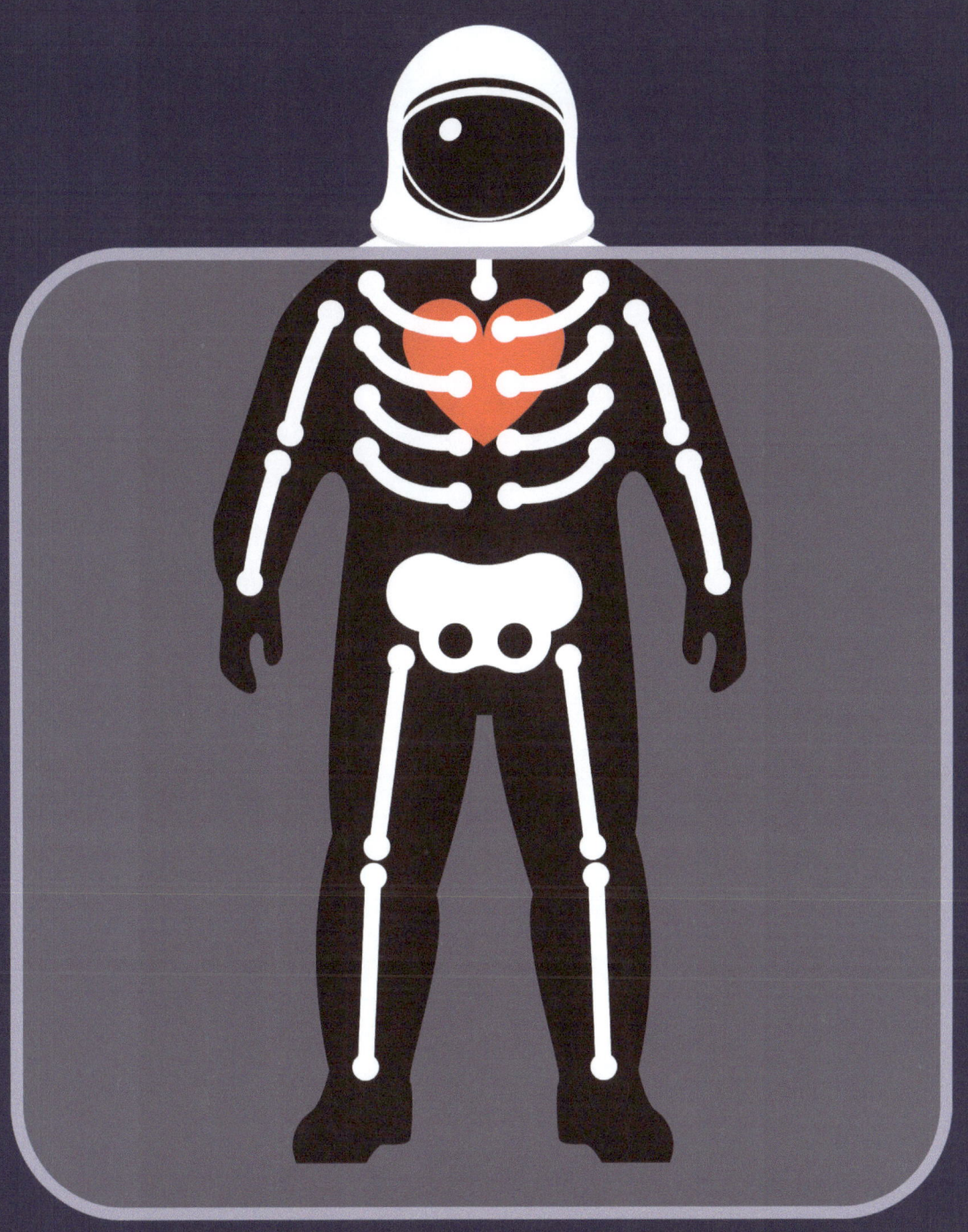

X-ray

Year - the length of time it takes for Earth to revolve around the sun. Earth rotates counterclockwise around the sun if viewed from North Pole, and clockwise if viewed from the South Pole.

Year

- January
- February
- March
- April
- May
- June
- July
- August
- September
- October
- November
- December

Zenith - a point directly overhead from an observer.

We hope you enjoyed this book, please consider leaving a review.

_ Thank You!

Copyright © 2022

All rights reserved. No part of this publication may be reproduced, distributed, or transmitted in any form or by any means, including photocopying, recording, or other electronic or mechanical method without prior written.

www.ingramcontent.com/pod-product-compliance
Lightning Source LLC
Chambersburg PA
CBHW051923210526
45473CB00006B/2112